GLOBAL WARMING THREATENS NATIONAL INTERESTS IN THE ARCTIC

There is international debate on the causes of global warming. Are the causes a result of human factors or is global warming a period in the earth's natural cycle? Although the answer is important, the impact of global warming, and how the United States reacts to it, will influence the nation's future status as a world leader. A United States' vital national interest is the protection and defense of its sovereign territory, but the possession of the natural resources within the sovereign territory is also an important national interest. Possession of natural resources is critical to economic growth and with the growth of many nations' economies will be an increase in demand for natural resources. The Arctic Ocean contains a vast amount of natural resources and the competition to claim those resources is on the rise, especially for oil and natural gas. Based on the world's current and future reliance on oil and natural gas, to include the United States, the possession of the oil and natural gas in the Arctic is a United States' important national interest.

Global warming has impacted the Arctic Ocean by significantly reducing the extent of the summer ice cover, allowing greater access to the region and to Alaska. As the earth's climate system continues to warm, access to the region will continue to increase. Greater access to the Arctic Ocean by United States' competitors and foes potentially threatens its national interests in the region. Fortunately, the United States has already begun to take action to protect its interests, but more can be done through continued international engagement, national initiatives, and specifically within the Department of Defense (DoD). The United States should continue to take actions to mitigate the potential threat to its national interests in the Arctic region caused by greater access to the region by its competitors and foes allowed by global warming.

United States' National Interests

On March 16, 2006, President George W. Bush released *The National Security Strategy of the United States of America*. The strategy identifies nine essential tasks that the United States must accomplish in order to meet the challenges it faces today. The task to "Transform America's National Security Institutions to meet the challenges and opportunities of the 21st century" includes the defense of the United States' sovereign territory.[1] Entry into the Unites States by terrorists, criminal organizations, or other enemies may threaten its government, infrastructure, and citizens. Therefore, the protection and defense of its sovereign territory is a United States' vital national interest.

Another task, "Ignite a new era of global economic growth through free markets and free trade," states that part of the United States' comprehensive energy strategy is to reduce its reliance on foreign energy sources.[2] It must increase the number of regions that provide it energy resources and the types of resources in order to ensure its energy security.[3] *The National Security Strategy of the United States of America* also states that "greater economic freedom is ultimately inseparable from political liberty" and that economic freedom empowers.[4] The possession of natural resources will enable a nation to achieve economic freedom if gathered and utilized properly. Therefore, the nations that have natural resources or acquire them through trade, and can effectively utilize them, are the nations most likely to possess the power and lead the world. Natural resources are a United States' important national interest.

Natural Resources of the Arctic Ocean and Alaska

One such area that contains vast natural resources is the Arctic Ocean. It is the smallest ocean in the world. It covers an area over 14 million square km (5.4 million square miles), has over 45,000 km (28,000 miles) of coastline, and is surrounded by the countries of the United States, Russia, Canada, Denmark, and Norway. It is predominantly covered by a perennial drifting ice cap that reaches from coastline to coastline in the winter and is surrounded by open water in the summer. The Northwest Passage and the Northern Sea Route are the two main waterways that connect the Atlantic Ocean with the Pacific Ocean. Under the Arctic Ocean, about 50 percent of the ocean floor is considered to be a continental shelf and the rest is a basin that is broken up by three ridges: Alpha Cordillera, Nansen Cordillera, and Lomonosov Ridge.[5] There are also vast tin, manganese, nickel, gold, platinum, and diamond reserves, and oil and natural gas reserves that could be close to 25 percent of the world's resources. In addition, there are vast quantities of fish stocks that migrate in its waters.[6] Although the smallest of the world's oceans, the Arctic Ocean has the potential to supply vast quantities of resources.

Along the coastline of the Arctic Ocean is the state of Alaska. In spite of the fact that the fishing and seafood, timber, mineral mining, and tourism industries are important to the economic success of Alaska, it is the oil and natural gas industry that earns the main revenue. It produces 25 percent of the United States' oil.[7] Alaska has the largest oil field in North America. It is called the Arctic Alaska Petroleum Province and it is located on the North Slope. It includes the Alpine oil field, National Petroleum Reserve in Alaska, Prudhoe Bay oil field, and the continental shelf located off of the northern coast. It is estimated to contain an additional 23 billion barrels of known recoverable crude oil and 52 trillion cubic feet of known recoverable natural gas. In addition, the area is estimated to contain both undiscovered, technically recoverable crude oil and natural gas; 24 billion barrels and 227 trillion cubic feet respectively.[8] In order to meet the United States' annual demand of 7.5 billion barrels of oil[9] and 23 trillion cubic feet of natural gas,[10] Alaska alone could theoretically provide the United States just over 6 years supply of oil and 12 years supply of natural gas. Therefore, it is a United States' important national interest to protect its natural resources that are available or potentially available in Alaska.

Competition for the Arctic Ocean

Besides the United States, there are other countries that have identified the potential and are exploiting the Arctic Ocean as an area of vast natural resources. In order to peacefully and legally exploit the natural resources of the world's oceans, the United Nations passed the United Nations Convention on the Law of the Sea (UNCLOS) on December 10, 1982, to address not only claims to the oceans' natural resources, but all other issues relating to the law of the sea. It declares that a coastal state's territory extends twelve nautical miles off of its coastline, but that it has exclusive economic rights to a 200 nautical mile exclusive economic zone (EEZ).[11] If the continental shelf extending from its territorial waters exceeds 200 nautical miles, the coastal state has the rights to the natural resources on and in that continental shelf out to 350 nautical miles. No other states are permitted to exploit the natural resources of that continental shelf without the authorization of the claiming coastal state. The claiming coastal state must submit charts or geographical coordinates of the continental shelf along with supporting scientific and technical data to the United Nations

Commission on the Limits of the Continental Shelf, requesting the establishment of the limits of the continental shelf per its data. If the Commission approves the request, the coastal state and the Secretary-General of the United Nations will be formally notified. If the coastal state does not agree with the decision of the Commission, it may resubmit a revised or new request.[12] If there is a conflict over the territory in question between states with opposite or adjacent coasts, an agreement will be made between the states per international law. If an agreement cannot be reached, the conflict will be settled via the International Tribunal for the Law of the Sea, the International Court of Justice, or an arbitral tribunal.[13] So, as defined by the UNCLOS, there are specific procedures and authorizations that must be addressed before a coastal state can peacefully and legally stake a claim to a continental shelf and the natural resources on and in it that extends beyond 200 nautical miles from the coastal state's territorial waters.

The United States, Canada, Russia, Norway and Denmark are exploiting the Arctic Ocean for its natural resources within their territorial waters and EEZs,[14] but some of them are also pursuing additional claims. In August 2007, Russia planted a titanium Russian flag on the Lomonosov Ridge, 4200 meters beneath the North Pole, bolstering their claim that the ridge is an extension of the continental shelf that extends out from their territorial waters and thus within their EEZ.[15] The Lomonosov Ridge is believed to contain vast reserves of oil, natural gas, and minerals that may be accessible in the future due to the decrease of the ice cap in the Arctic Ocean. Since Russia ratified the UNCLOS in 1997, it tried to claim the Lomonosov Ridge in 2001, but the United Nations needed more data before it could make a decision. Russia continues to study the ridge and plans to provide the needed evidence in 2009 to the United

Nations to stake their claim.[16] Following Russia's lead, Danish and Canadian scientists are also studying the Lomonosov Ridge hoping to collect enough evidence to also claim the ridge as part of the continental shelf off the coasts of Canada and Greenland and within their EEZs.[17] Even the United States has been conducting sea bed geological surveys in the Arctic. This past fall, in conjunction with Canada, it conducted a scientific exploration of the Arctic sea floor to map the outer limits of its continental shelf in accordance with the UNCLOS criteria.[18] Thus, the United States is also trying to extend its EEZ. The competition for natural resources in the Arctic Ocean has begun.

Supply and Demand for Oil and Natural Gas

It can be argued that to reduce the price of oil in the United States, additional drilling in the United States is required. Since the summer of 2008, and even before then, when the price of gasoline was on the rise and oil was breaking the cost-per-barrel records, the United States has been focused on the supply and demand of oil. President Bush, Senator John McCain and many Americans believed that additional drilling in the United States, either offshore or in Alaska, would provide some relief to the rising oil prices. Opponents to the idea claim that it would take too long to establish before any results would be realized. Organization of Petroleum Exporting Countries (OPEC) blamed the high prices on speculators and would not adjust their production.[19] Fortunately, the fall of 2008 has seen the reduction in oil costs. The reduction in the cost-per-barrel is due to large decreases in demand resulting in large inventories of oil. As oil companies adjust to the changes in demand, striving to balance supply with demand, the price of oil will again increase.[20] Adjustments in the price will continue, but not from a lack of oil reserves. The speculations of future demand and the quantities made available to consumers by oil companies will be the determining factors. Therefore, there is no urgent need for additional drilling in the United States, either offshore or in Alaska, at this time.

Although there isn't an urgent need for additional drilling now, there may not be a choice in the next twenty-five years. With the exception of Eastern Europe and the former USSR, the demand for oil by the other regions of the world has consistently increased over the past twenty-plus years.[21] Similarly, the demand for natural gas has also increased.[22] Over the next twenty-five years, the demand for oil and natural gas will continue to increase, especially in the transportation sector, but also in the world's emerging economies that will modernize their transportation systems to move products and raw materials.[23] Although new technology will gain efficiencies in motor vehicles and airplanes, the projected growth in demand for transportation of people and things due to population growth and increases in the Gross Domestic Product will outpace the technological improvements in transportation system efficiencies.[24] Since the United States relies heavily on its transportation sector to move people and things, its demand for energy will also increase. Unfortunately, transitioning to another type of fuel or energy source will be difficult. The transition will require large changes in the motor vehicle fleet, fueling stations, and the fuel distribution infrastructure.[25] Consequently, the United States' demand for oil and natural gas will also increase in the foreseeable future.

To meet today's demand for oil, the United States obtains over 40 percent of its oil requirements from resources within the United States.[26] The main oil reserves are located in Texas, Alaska, California, Wyoming, and New Mexico, and in the Federal

Offshore fields in the Gulf of Mexico and the Pacific Ocean.[27] Over 80 percent of the United States' requirement for natural gas is obtained within the United States,[28] mainly in Texas, Wyoming, New Mexico, Oklahoma, Colorado, and the Gulf of Mexico Federal Offshore field.[29] Currently, the United States continues to study and search for new sources of oil and natural gas, but in the meantime, it imports its deficit.

The United States imports 58 percent of the petroleum it consumes.[30] It imports the majority of its crude oil and petroleum from Canada, Saudi Arabia, Mexico, and Venezuela.[31] Fortunately, its suppliers have as much to gain from the trade as the United States. Canada has the second largest oil reserve in the world and 99 percent of its oil exports go to the United States.[32] Saudi Arabia has the largest oil reserves in the world and is the world's leading oil producer and exporter. The United States shares many common interests with Saudi Arabia such as regional stability, counter-terrorism, and the resolution of the Israeli-Palestinian conflict, and it considers them an ally.[33] Mexico is the third largest supplier of oil to the United States.[34] They share many common interests from migration and trafficking across a common border to counter-terrorism that requires a strong relationship between the two to overcome their common challenges. Lastly, Venezuela and the United States are mutually dependent. Although only 11 percent of United States' crude oil is imported from Venezuela, it is 60 percent of Venezuela's crude oil exports. It also has five state owned refineries in the United States. Venezuela wants to lessen its dependence on the United States by exporting its oil to more countries, but the economic benefits of dealing with the United States will make it difficult and slow to make the change.[35] By importing the majority of its crude oil and petroleum from either neighboring nations that share its borders and interests, to an ally that shares common interests, to a nation that relies heavily on its oil exports to the United States for economic success, the United States is not, and will not be in the foreseeable future, at risk of insufficient access to crude oil and petroleum resources.

The United States is the largest producer of natural gas in the Western hemisphere,[36] consequently, only 19 percent of the natural gas consumed by the United States is imported.[37] It imports the majority of its natural gas from Canada, Trinidad, Egypt, Mexico, and Nigeria.[38] Canada is the second largest producer of natural gas in the Western hemisphere and provides the United States 86 percent of its natural gas imports.[39] As with crude oil and petroleum imports, by importing the majority of its natural gas from Canada, which is only a fifth of its requirement, the United States is not, and will not be in the foreseeable future, at risk of insufficient access to natural gas resources.

As mentioned before, the demand for liquid fuel will consistently increase as the transportation sector grows, especially in the emerging economies of the world. The Energy Information Administration's *International Energy Outlook 2008* reports that liquid fuels consumption will increase from 84 million barrels per day in 2005 to 113 million barrels per day in 2030. Although the increase is nearly 30 million barrels a day, the report also states that through the continued production of conventional liquid fuels in addition to increased production of unconventional liquid fuels, the demand will be fulfilled out to the year 2030. The increase in unconventional liquid fuels includes the production of extra-heavy oil, biofuels, coal-to-liquids (CTL), and gas-to-liquids.[40] In the year 2030, the Energy Information Administration expects that nearly twenty percent of the world's liquid fuel supply will be provided by unconventional sources as long as oil prices remain high, meaning prices that are comparative to the oil prices that the world experienced in the first eight months of 2008. The high oil prices will allow for the development of unconventional resources and the more efficient recovery of conventional resources.[41] In the event that oil prices remain low, approximately seventy dollars per barrel of oil and below, the world demand will also be met. The OPEC and non-OPEC countries will increase their conventional liquid fuel production, but due to the lack of investment dollars for the research and development of new conventional liquid fuel resources, it will lead to the eventual exhaustion of their supply. The production of unconventional liquid fuel will decrease due to the lack of incentive to conserve fuel and the investment dollars to produce it economically.[42]

Domestically, the United States' oil imports are expected to decrease by 2030 even though demand will increase.[43] Oil production in the Gulf of Mexico will increase by a million barrels per day by 2019 and then steadily decline due to exhaustion of the oil fields. Conversely, United States' production of biofuels is projected to increase by a million barrels per day by the year 2030, half of the world's biofuels production. It will also increase its CTL production by a quarter million barrels per day by 2030. The increase in the United States' production of unconventional liquid fuels is based on the expectation of high oil prices to cover the costs of development.[44] Per *The National Security Strategy of the United States of America* and its comprehensive energy strategy, the increase in the production of unconventional liquid fuels is the direction that the United States wants to advance - to reduce its reliance on foreign energy sources.

Although the *International Energy Outlook 2008* has optimistic expectations to meet the future demand for liquid fuels, the expectations are based on many variables that must be realized in order to meet the demand in 2030 and beyond. The greatest variable is the price of oil. The price of oil is dependent upon the demand for liquid fuels, production costs for conventional liquid fuels, investment and production behavior, and the cost and availability of unconventional liquid fuels.[45] In addition, the current resource, technological, and infrastructure challenges impede the use of biofuels and CTL[46] as an increased supplement or alternative to conventional liquid fuels, thus possibly requiring the United States to remain dependent on oil. As the United States continues to exhaust its known conventional liquid fuel reserves, it may depend more upon imports to meet its demand. Without sufficient unconventional liquid fuels to meet the increasing demand for liquid fuels, the United States will continue to rely on foreign energy sources, countering its comprehensive energy strategy. Access to the proved reserves and undiscovered recoverable resources of oil and natural gas in the Arctic Ocean is a United States' important national interest.

Impact of Illegal, Unregulated and Unreported Fishing (IUU)

> According to the United Nations Food and Agriculture Organization, 75% of the world's major marine fish stocks are fully exploited, overexploited, or significantly depleted. As the world population continues to grow and world food demand increases, fishing pressure on high seas, migratory and straddling fish stocks will intensify.[47]

IUU is a global problem. It causes serious economic, social and environmental challenges. Annual IUU costs have been estimated at 15.5 billion dollars globally.[48] The fishing industries that have been affected are the Alaska pollock, Pacific cod, safron cod, Pacific herring, and several species of salmon, redfish, halibut, flounder, squid, crab and shrimp.[49] Some of the major IUU impacts and challenges include the reduction in potential employment that local and locally based fleets provide; the reduction in resources that will result in decreased revenues for other companies providing fishing services; damage to inshore prawn fishing areas and mangrove areas resulting in decreased income for coastal fishing communities;[50] and the rapid, undocumented reduction in specific fish populations that could negatively affect the ecosystem. In addition, climate change already challenges scientists in predicting the future of fish populations and IUU only adds to their challenge.[51] The United States has borne some of that cost from IUU in its EEZ in the Western Bering Sea. The actual amount is unknown, but IUU negatively impacts the United States' access to its natural resource.

Currently, the United States Coast Guard (USCG), as one of its maritime security responsibilities, protects living marine resources through the detection and deterrence of illegal fishing activities. One of its areas of responsibility is the EEZ that surrounds Alaska. South of the Bering Strait, it predominantly has focused its efforts in the Western Bering Sea along the United States - Russia Maritime Boundary Line.[52] There has been little focus in the Arctic Ocean north of the Bering Strait since there has been no commercial fishing in that area.[53] With the receding summer ice cover caused by global warming allowing access to the Arctic Ocean, IUU will most likely be another challenge for the USCG in the United States' EEZ in the Arctic Ocean.

USCG Initiatives North of the Arctic Circle

Although IUU has not been a predominant focus for the USCG in the Arctic Ocean, it has focused its operations on other challenges. The USCG's 17th District is responsible for Alaska and the Arctic region. Its mission is to "serve and safeguard the public, protect the environment and its resources, and defend the Nation's interests in the Alaskan maritime region."[54] In the summer of 2008, besides conducting its regular missions of maritime law enforcement, maritime safety, search and rescue, maintaining aids to navigation, marine environmental response, homeland security and defense, and ice operations predominantly south of the Bering Straits, the 17th District also conducted operations and exercises north of the Arctic Circle. It conducted C-130 air patrols out of Nome, Alaska in order to conduct Arctic domain awareness and provide scientists an airborne platform for their research. The *USCGC Healy* deployed three times to the Arctic in support of the National Science Foundation. The *USCGC Spar* conducted Waterways Analysis and Management System assessment of the Alaskan North Slope. The USCG planned a joint search and rescue exercise with the other Arctic nations' search and rescue assets to build partnerships and test capabilities; it deployed two helicopters and two 25' small boats via C-130 to Barrow, Alaska to liaise with the local community and conduct boating safety training; and it deployed another detachment to Prudhoe Bay, Alaska to not only liaise with the local community, but also to conduct an Arctic security exercise. The USCG is preparing to also conduct all of its missions in northern Alaska that it now conducts in southern Alaska. It is the change in the Arctic region that is expanding the USCG's mission to the north.[55]

Impact of Global Warming on the Arctic Ocean

The Intergovernmental Panel on Climate Change (IPCC), created in 1995, has stated that it is unequivocal (90 percent certain) that the earth's climate system is warming.[56] The CNA Corporation, a nonprofit national security analysis organization, convened a panel of flag and general officers and national security experts to study the impact on the United States' national security by global climate change.[57] According to their study, the impact on the world is that it will experience extreme weather events, rise in the sea level, droughts, flooding, retreating glaciers, migration of people to lesser affected areas, and an increase in the spreading of disease.[58] The greatest effect of global warming has been in the Arctic and the melting of the ice cap.[59] The increase in global temperatures has caused a significant reduction in the extent of the summer Arctic Ocean ice cover. In 2007, the extent of the Arctic Ocean ice cover was 39 percent below the long-term average from 1979 to 2000 and believed to be 50 percent lower than the annual minimums from the 1950s to the 1970s. It wasn't until 1979 that accurate ice cover could be measured utilizing satellite-based passive microwave images.[60] Based on the fact that the summer Arctic Ocean ice cover is declining every year, there are many predictions as to when the Arctic Ocean will be ice free during the summer months. It could be as early as 2013 or as late as the year 2100. The National Snow and Ice Data Center predicts it will be closer to the year 2013 than 2100.[61]

Although the Arctic Ocean was not completely ice free in the summer months, the area off the north coast of Alaska and Canada in the Beaufort Sea was ice free during the months of August and September 2008. In fact, when comparing satellite images of Arctic Ocean ice cover, that has been the case most years dating back to 1979.[62] However, this is the first year that the Northwest Passage has been clear of enough ice to allow for a commercial ship to transit through. The Canadian Coast Guard confirmed that a commercial cargo ship transited the Northwest Passage from Montreal, Quebec to communities in western Nunavut, Canada in the fall of 2008.[63] Although it didn't transit all of the way through the Northwest Passage from the Atlantic Ocean to the Pacific Ocean, it is the first since the *SS Manhattan*, a specially reinforced supertanker, and the *John A Macdonald*, a Canadian icebreaker, made the journey in 1969. The journey was made to test the Northwest Passage as an alternative to building the Alaska Pipeline, but it was determined that it was not economical, so the Alaska Pipeline was built.[64] With the melting trend of the Arctic Ocean ice cover and the first of probably many successful attempts to journey the Northwest Passage, the maritime traffic in the Northwest Passage, the Northern Sea Route, and the Arctic Ocean is sure to increase as access to the Arctic Ocean increases. Consequently, it may be economically more feasible to transit via the Arctic Ocean as a route through the Arctic Ocean from China to Europe would save 5,000 miles when compared to a route passing through the Panama Canal or Suez Canal.[65]

Potential Risk to United States' National Interests

The opening of the Northwest Passage and the Northern Sea Route and greater access to the Arctic Ocean may someday be economically beneficial, but it also potentially puts the United States' national interests – natural resources and sovereignty – at greater risk. The opening of the Arctic Ocean allows other nations greater opportunity to seek and claim areas abundant with natural resources such as Russia's claim to the Lomonosov Ridge, and the Danish and Canadian scientists who are studying the Arctic Ocean floor hoping to collect enough evidence to extend their EEZs past the 200 mile limit. The opportunity for IUU in international waters and the United States' EEZ will increase as fishing vessels will be able to access fish stocks as they migrate north with the receding ice cover. The opportunity for terrorist and criminal organizations or other enemies to access the United States via remote locations now more accessible in northern Alaska may put not only the oil and natural gas resources and production facilities in the Arctic Alaska Petroleum Province to include the Alaskan Pipeline at risk, but also key infrastructure within the continental United States. Once a terrorist or members of a criminal organization gain access to sovereign United States territory, a passport is not required to commercially fly from Alaska to the continental United States. They could potentially travel unhindered to almost anywhere in the United States allowing greater access to key infrastructure.

Greater access to the Arctic Ocean may also result in not only a greater number of commercial vessels, but also recreational vessels as tourism will also increase. Therefore, the requirement for maritime law enforcement and safety, search and rescue, and environmental response may also increase. Lastly, the opening of the Arctic Ocean will provide another area for nations to conduct air and naval operations and potentially challenge United States' sovereignty. Since August 2007, Russian strategic bombers have flown patrol flights over the Arctic to provide a routine Russian presence in what they describe as a highly strategic area, but also to improve aircrew proficiency in flying and navigating over the vast and featureless Arctic region. For the first time in seventeen years, the Russian Navy resumed its regular naval presence in the Arctic. In June and July 2008, the Russian Udaloy class anti-submarine ship *Severomorsk* and the Slava class missile cruiser *Marshal Ustinov* deployed to the Arctic.[66] Consequently, greater access to the Arctic Ocean due to the decreased ice cover and the potential opening of the Northwest Passage and the Northern Sea Route may pose a threat to the United States' national interests in that area.

Recommended United States' Actions

In order to reduce the potential threat to the United States' national interests by the impact of global warming on the Arctic Ocean, the United States should continue to act internationally, nationally, and within the DoD. Although access to the Arctic Ocean is expected to gradually increase in the future and to be ice free sometime between now and the year 2100, there are additional actions that can be and should be taken now to better prepare the United States for the future.

On the international stage, the United States has already been actively involved in trying to understand and in addressing the impacts of global warming. It has initiated or participated in dozens of partnerships that include many programs, from developing transformational low-carbon technologies to improving observation systems. Examples include the Asia-Pacific Partnership on Clean Development and Climate that focuses on reducing greenhouse gas emissions and increasing economic development and the United Nations Framework Convention on Climate Change that identified many actions that can be taken now to reduce emissions.[67] The United States is also an active participant on the IPCC, the United Nations Environment Program, and other international conventions, programs, and organizations that address climate change.[68] Since 2001, the United States has spent 45 billion dollars on programs in science and technology to better understand and address climate change.[69]

In addressing the challenges with the Arctic region, the United States is a member of the Arctic Council. The Arctic Council is a high-level diplomatic forum composed of members from the eight Arctic states and the only major diplomatic forum devoted to Arctic issues. The issues the council addresses include: national security and homeland security, international governance, extended continental shelf and boundary issues, international scientific cooperation, shipping, economic and energy issues, environmental protection, and conservation of natural resources.[70] In May 2008, the Arctic states issued the *Ilulissat Declaration* that reaffirmed their cooperation and commitment. The first issue addressed in the *Ilulissat Declaration* was the safety of life at sea. With increasing tourism and human presence in the Arctic Ocean, the *Ilulissat Declaration* identified the need to cooperate on search and rescue operations, to respond to accidents and environmental emergencies, and to maintain regular and open lines of communication.[71] The United States has found that the Council is a valuable forum and is committed to continuing its participation and support.[72]

The United States is already heavily engaged with its Arctic partners via the Arctic Council and the United Nations in addressing the impact of global warming in the Arctic region. It must continue to focus on the issues that potentially threaten its national interests, its rightful access and claims to natural resources, especially oil and natural gas. It must continue to partner with the other Arctic states in the scientific research and technological development in accessing those resources, and ensure via the United Nations that territorial claims are justified with scientific data. It must also continue its efforts with the other nations in the defeat of IUU by coordinating law enforcement actions, information sharing, and commercial fisheries management to include tighter port controls on the import of fish. Lastly, it must not only continue to engage the other Arctic states in maritime safety and environmental emergency response issues, but also expand its capabilities by conducting and or participating in combined search and rescue and environmental emergency response exercises and operations. The Arctic

Ocean is a large area and only getting larger as the sea ice cover recedes. It will take consistent engagement, cooperation, and compromise to be successful in the region.

Nationally, the most immediate action that can be taken by the United States to reduce the potential threat to its national interests in the Arctic region is to ratify the UNCLOS. The primary functions of UNCLOS "are to define maritime zones, protect the marine environment, preserve freedom of navigation, allocate rights to resources, and establish certain necessary guidelines for businesses that depend on the sea for various purposes."[73] The primary functions provide the legal framework and support the United States' efforts to protect its national interests, especially in the Arctic region.[74] However, there has been reluctance by a portion of the Senate to ratify the UNCLOS. Their objection is that, according to the UNCLOS, only a United Nations' court or tribunal can resolve maritime issues involving preservation, research, navigation, fisheries, and marine environmental protection. Consequently, the UNCLOS could restrict United States' autonomy.[75] Additionally, those opposed to the UNCLOS state that it would allow the United Nations to levy taxes, allow an international tribunal to question the authority of the United States Navy, impair intelligence and submarine activities and require technology sharing. They also claim that the United States does not require the UNCLOS because it can threaten or use force to protect its navigational rights or rely on customary international law.[76] According to John B. Bellinger III, Legal Advisor for the United States Department of State, the opposition's arguments are "inaccurate, outdated, or incomplete."[77]

Currently, 152 nations have ratified the treaty. The UNCLOS was negotiated in 1982 to replace the 1958 High Seas Treaty, and in 1983, President Reagan directed the

United States to comply with all of its provisions except for Part XI, which addressed deep seabed mining. In 1994, Part XI of the UNCLOS was officially modified to address the United States' concerns.[78] It wasn't until the Bush Administration took office and reviewed all of the un-ratified treaties that action was again taken to ratify the UNCLOS. In February 2004, the Senate Foreign Relations Committee voted unanimously to approve the UNCLOS, but due to it being an election year, it was never put before the full Senate for a vote to ratify. Consequently, the UNCLOS dropped off of the Senate agenda until 2007 when Russia planted its titanium flag beneath the North Pole bolstering their claim to the Lomonosov Ridge. In October 2007, the Senate Foreign Relations Committee voted 17 to 4 to approve the UNCLOS, but once again it was not put before the full Senate for a vote. Based on the opposition's intent to draw out the argument, the Senate Majority Leader decided to postpone the full Senate vote to ratify the UNCLOS.[79]

Ratification of the UNCLOS has the support of many of the United States' senior leaders, organizations, and businesses to include President Bush, the Chairman of the Joint Chiefs of Staff, the Commandant of the USCG, former Secretary of State Condoleezza Rice, the Natural Resources Defense Council, the Western Pacific Fisheries Management Council, and many more.[80] By ratifying the UNCLOS, the United States will have the legal authority of the convention to protect its national interests. It will be in a legitimate position to defend against any hostile recommendations to change the UNCLOS and support positive initiatives. It will have a voting seat on the Council of the International Seabed Authority that makes decisions on deep seabed mining issues. It will announce to the world that it is a committed partner in managing the world's oceans and natural resources. Lastly, ratification of the UNCLOS would not change the United States' foreign or domestic policy.[81] Therefore, the United States should expeditiously ratify the UNCLOS.

The Department of Homeland Security, specifically the USCG, has identified the increase in activities in the Arctic region from the Arctic states' efforts to claim and gain access to its natural resources to commerce, tourism and research.[82] In its Strategic Priorities for Fiscal Year 2009 outlined in the *U.S. Coast Guard Posture Statement*, the USCG has already begun to focus its efforts in increasing its operations in the region to protect United Sates sovereignty and its national interests.[83] The USCG needs to continue focusing on the Arctic region and develop a long range plan that addresses the operational, fiscal and support requirements of the 17th District as its operational requirements increase due to the receding ice cover in the Arctic Ocean.

Lastly, the United States needs to continue to focus on developing alternate sources of energy and the infrastructure to support it, reducing conventional liquid fuel consumption, and discovering and accessing additional oil and natural gas resources. Under the Bush Administration, many initiatives have been started to reduce the United States' reliance on foreign energy sources, such as the Major Economies Meetings on Energy Security and Climate Change, the International Partnership for the Hydrogen Economy, the Washington International Renewable Energy Conference, the Energy Independence and Security Act, Climate VISION, the Green Power Partnership, the SmartWay Transport Partnership, and many others that mainly address greenhouse gas emissions.[84]

Although the Bush Administration has taken action to reduce the United States reliance on foreign energy sources, it has been criticized internationally for opposing similar initiatives such as the Kyoto Protocol. The Kyoto Protocol is a United Nations' treaty that calls for the industrialized nations to reduce their greenhouse gas emissions over a five year period, from 2008 to 2012, to below 1990 levels. International criticism, specifically by the United Nations, is based on the fact that the United States emits 24 percent of the global greenhouse gas, twice as much as the second leading nation, China.[85] The United States opposes the treaty claiming that there is not enough adequate scientific data that supports the successful achievement of the treaty's goals, and support of the treaty would raise energy costs and cause severe economic damage.[86] Therefore, the new Obama Administration needs to continue, improve upon and support the current initiatives in order to achieve the United States' comprehensive energy strategy endstate, but also maintain engagement with the United Nations and the international community in resolving the greenhouse gas emission issues and the impact of global warming.

With the increased surface presence of the Russian Navy after seventeen years of ignoring the Arctic region, the DoD in coordination with the Department of Homeland Security should conduct a risk analysis on increasing its future presence in the Arctic region to deter other nations and organizations from threatening United States' sovereignty. Although the Arctic Ocean is only open for a couple months out of the year, a naval presence or overflights of the region tells the world that the United States has national interests in the region, could possibly deter hostile infractions upon its sovereignty, allows the United States Navy and Air Force to exercise their navigational rights, and provides training to its military forces in operating in that type of environment. Currently there are no indications of a hostile threat to United States' sovereignty or its natural resources in the region, but as the competition for natural resources increases, and the region provides greater accessibility to the United States and its critical infrastructure, attempts to exploit the vulnerability of the Arctic region may be inevitable.

Due to the length of time required to modify the structure of the military and steep competition for governmental funding, the DoD needs to identify its service requirements to provide a military presence in the Arctic region. From the risk analysis and a thorough study of the requirements for a military presence in the region; planning, budgeting, personnel management, recruiting, and acquisition or modification of equipment and weapons systems should begin as soon as the requirements are identified. The requirements may only require the diversion of a carrier battle group to the Arctic from its regularly scheduled overseas deployment or changes made to the flight paths of aircraft already operating in the region. At a minimum, the services should require their future acquisitions of military equipment and weapons systems to be able to also operate in the Arctic environmental conditions. In addition, modifications should be made as required to current equipment and weapons systems to operate in those conditions. In order to defend the nation, the military services must be able to operate in any clime and place.

As "the largest oil consuming government body in the United States and in the world,"[87] the DoD needs to continue to research and develop alternate energy sources. The advancement that has been made with the testing of synthetic fuels in the United States Air Force's B-52s is the direction that the DoD needs to advance. The use of a domestically sourced synthetic fuel to power its fleet of aircraft is an Air Force goal and adheres to the United States' comprehensive energy strategy.[88] It only makes sense that the United States would not want its military forces reliant upon foreign energy sources.

Conclusion

Global warming has impacted the Arctic Ocean by significantly reducing the extent of the summer Arctic Ocean ice cover allowing greater access to the region and to Alaska. Greater access to the Arctic Ocean potentially threatens United States' national interests in the region. Access to the United States' natural resources, specifically oil and natural gas, and to its territory and critical infrastructure must be protected. The United States must continue to engage the United Nations and the Arctic Council on issues that potentially threaten its national interests. The new Obama Administration should continue, improve upon and support the international and domestic climate change initiatives that were begun under President Bush to include the nation's comprehensive energy strategy. In addition, ratification of the UNCLOS should be an immediate priority for the new administration. The Department of Homeland Security and DoD must continue to study the requirements and plan for increased operations in the Arctic region, and DoD must continue to research and develop new and alternate energy sources for their forces. Global warming is a difficult challenge and its impact can only be mitigated by a holistic government approach.

Endnotes

[1] George W. Bush, *The National Security Strategy of the United States of America* (Washington, DC: The White House, March 2006), 43.

2 Ibid., 1.

3 Ibid., 28-29.

4 Ibid., 25.

5 "Arctic Ocean," December 4, 2008, linked from *Central Intelligence Agency Home Page* at "The World Factbook," https://www.cia.gov/library/publications/the-world-factbook/geos/xq.html (accessed December 8, 2008).

6 Victor Yasmann, "Russia: Race to the North Pole," July 27, 2007, linked from *GlobalSecurity.org Home Page* at "Military," http://www.globalsecurity.org/military/library/news/2007/07/mil-070727-rferl02.htm (accessed December 1, 2008).

7 Alaska Railroad, "Alaska Facts," July 22, 2004, http://www.alaskarails.org/AK-facts.html (accessed December 8, 2008).

8 David W. Houseknecht and Kenneth J. Bird, "Oil and Gas Resources of the Arctic Alaska Petroleum Province," *U.S. Geological Survey Professional Paper 1732–A*, 2006, http://pubs.usgs.gov/pp/pp1732/pp1732a/pp1732a.pdf (accessed December 9, 2008).

9 Energy Information Administration, "Product Supplied for Crude Oil and Petroleum," July 28, 2008, http://tonto.eia.doe.gov/dnav/pet/pet_cons_psup_dc_nus_mbblpd_a.htm (accessed December 7, 2008).

10 Energy Information Administration, "Natural Gas Consumption by End Use," November 26, 2008, http://tonto.eia.doe.gov/dnav/ng/ng_cons_sum_dcu_nus_a.htm (accessed December 9, 2008).

11 "Part V. Exclusive Economic Zone," linked from *United Nations Convention on the Law of the Sea*, December 10, 1982, at "Part V," http://www.un.org/Depts/los/convention_agreements/texts/unclos/closindx.htm (accessed December 13, 2008).

12 "Annex II. Commission of the Limits of the Continental Shelf," linked from *United Nations Convention on the Law of the Sea*, December 10, 1982, at "Annex II," http://www.un.org/Depts/los/convention_agreements/texts/unclos/closindx.htm (accessed December 13, 2008).

13 "Part XV. Settlement of Disputes," linked from *United Nations Convention on the Law of the Sea*, December 10, 1982, at "Part XV," http://www.un.org/Depts/los/convention_agreements/texts/unclos/closindx.htm (accessed December 13, 2008).

14 "Russian Explorers Plant Flag on North Pole Seabed," August 2, 2007, linked from *GlobalSecurity.org Home Page* at "Military," http://www.globalsecurity.org/military/library/news/2007/08/mil-070802-rferl02.htm (accessed December 1, 2008).

15 Ibid.

16 "Russia to Continue Arctic Shelf Research," March 27, 2008, linked from *GlobalSecurity.org Home Page* at "Military," http://www.globalsecurity.org/military/library/news/2008/03/mil-080327-rianovosti04.htm (accessed December 1, 2008).

[17] Yasmann, "Russia: Race."

[18] Environment News Service, "U.S. and Canada Collaborate on Arctic Sea Bed Mapping," September 3, 2008, http://www.ens-newswire.com/ens/sep2008/2008-09-03-02.asp (accessed December 23, 2008).

[19] Richard Valdmanis and Matthew Robinson, "Most Americans support more U.S. oil drilling: poll," June 18, 2008, linked from *Reuters Home Page* at "Politics," http://www.reuters.com/article/politicsNews/idUSN1844581020080618 (accessed December 7, 2008).

[20] Ann Davis, Ben Casselman and Carolyn Cui, "Oil's Slide Set to Leave Dark Trail," December 5, 2008, linked from *The Wall Street Journal Home Page* at "Economy," http://online.wsj.com/article/SB122843588512681303.html?mod=googlenews_wsj (accessed December 7, 2008).

[21] Energy Information Administration, "World Oil Demand by Region," http://www.eia.doe.gov/pub/oil_gas/petroleum/analysis_publications/oil_market_basics/dem_image_consumption.htm (accessed December 5, 2008).

[22] "Highlights," June 2008, linked from *Energy Information Administration* at "International Energy Outlook 2008," http://www.eia.doe.gov/oiaf/ieo/highlights.html (accessed December 2, 2008).

[23] Ibid.

[24] Energy Information Administration, "Petroleum Products: Consumption," September 2008, http://www.eia.doe.gov/neic/infosheets/petroleumproductsconsumption.html (accessed December 1, 2008).

[25] Energy Information Administration, "Highlights."

[26] Energy Information Administration, *Annual Energy Review 2007* (Washington, DC: U.S. Department of Energy, June 2008), 137.

[27] Energy Information Administration, "Petroleum Reserves," February 2008, http://www.eia.doe.gov/neic/infosheets/petroleumreserves.html (accessed December 7, 2008).

[28] Energy Information Administration, "Natural Gas Supply," October 2008, http://www.eia.doe.gov/neic/infosheets/natgassupply.html (accessed December 2, 2008).

[29] Energy Information Administration, "U.S. Crude Oil, Natural Gas, and Natural Gas Liquids Reserves 2006 Annual Report," November 2007, http://www.eia.doe.gov/pub/oil_gas/natural_gas/data_publications/crude_oil_natural_gas_reserves/current/pdf/arr.pdf#page=41 (accessed December 4, 2008), 31.

[30] Energy Information Administration, *Annual Energy Review 2007*.

[31] Energy Information Administration, "Crude Oil and Total Petroleum Imports Top 15 Countries," November 26, 2008, http://www.eia.doe.gov/pub/oil_gas/petroleum/data_publications/company_level_imports/current/import.html (accessed December 2, 2008).

[32] Energy Information Administration, "Country Analysis Briefs Canada," May 2008, http://www.eia.doe.gov/emeu/cabs/Canada/pdf.pdf (accessed December 4, 2008).

[33] U.S. Department of State, "Background Note: Saudi Arabia," February 2008, http://www.state.gov/r/pa/ei/bgn/3584.htm (accessed December 4, 2008).

[34] U.S. Department of State, "Background Note: Mexico," November 2008, http://www.state.gov/r/pa/ei/bgn/35749.htm#relations (accessed December 4, 2008).

[35] Cesar J. Alvarez and Stephanie Hanson, "Venezuela's Oil-Based Economy," June 27, 2008, http://www.cfr.org/publication/12089/venezuelas_oilbased_economy.html#7 (accessed December 4, 2008).

[36] Energy Information Administration, "Country Analysis Briefs Canada."

[37] Energy Information Administration, "Natural Gas Supply."

[38] Energy Information Administration, "U.S. Natural Gas Imports by Country," November 26, 2008, http://tonto.eia.doe.gov/dnav/ng/ng_move_impc_s1_m.htm (accessed December 2, 2008).

[39] Energy Information Administration, "Country Analysis Briefs Canada."

[40] "Chapter 2. Liquid Fuels," June 2008, linked from *Energy Information Administration Home Page* at "International Energy Outlook," http://www.eia.doe.gov/oiaf/ieo/pdf/liquid_fuels.pdf (accessed December 23, 2008).

[41] Ibid.

[42] Ibid.

[43] "Annual Energy Outlook 2009 Early Release Summary Presentation," December 2008, linked from *Energy Information Administration Home Page* at "Annual Energy Outlook," http://www.eia.doe.gov/oiaf/aeo/aeo2009_presentation.html (accessed December 23, 2008).

[44] Energy Information Administration, "Chapter 2. Liquid Fuels."

[45] Ibid.

[46] Energy Information Administration, "Nonconventional Liquid Fuels," 2006, http://www.eia.doe.gov/oiaf/aeo/otheranalysis/aeo_2006analysispapers/nlf.html (accessed December 23, 2008).

[47] United States Coast Guard, "OCEAN GUARDIAN, U.S. Coast Guard Fisheries Enforcement Strategic Plan," September 20, 2004, http://www.uscg.mil/hq/cg5/cg531/LMR.asp (accessed December 26, 2008).

⁴⁸ World Wildlife Fund, *Illegal Fishing in Arctic Waters* (Oslo, Norway: WWF International Arctic Programme, April 2008), v.

⁴⁹ Ibid., 18.

⁵⁰ Ibid., 24.

⁵¹ Ibid., 6.

⁵² United States Coast Guard, "OCEAN GUARDIAN."

⁵³ North Pacific Fishery Management Council, "Arctic Fishery Management," http://alaskafisheries.noaa.gov/npfmc/current_issues/Arctic/arctic.htm (accessed December 26, 2008).

⁵⁴ *United States Coast Guard 17th District*, http://www.uscg.mil/d17/ (accessed December 26, 2008).

⁵⁵ United States Coast Guard 17th District, "Coast Guard Arctic Operations: An Overview," http://www.uscg.mil/d17/ArcticOverview.pdf (accessed December 26, 2008).

⁵⁶ Carolyn Pumphrey, "Introduction," in *Global Climate: National Security Implications*, ed. Carolyn Pumphrey (n.p.: Strategic Studies Institute, May 2008), 2.

⁵⁷ The CNA Corporation, "National Security and the Threat of Climate Change," 2007, http://securityandclimate.cna.org/ (accessed October 25, 2008), 9.

⁵⁸ Ibid., 6.

⁵⁹ Ibid., 38.

⁶⁰ National Oceanic and Atmospheric Administration, "Arctic Report Card 2008," October 14, 2008, http://www.arctic.noaa.gov/reportcard/seaice.html (accessed December 26, 2008).

⁶¹ "Will the ice at the North Pole melt?" linked from *The National Snow and Ice Data Center Home Page* at "Arctic Sea News and Analysis Frequently Asked Questions," http://www.nsidc.org/arcticseaicenews/faq.html (accessed December 26, 2008).

⁶² The National Snow and Ice Data Center, "Arctic Sea News and Analysis," http://www.nsidc.org/arcticseaicenews/ (accessed December 26, 2008).

⁶³ "1st commercial ship sails through Northwest Passage," November 28, 2008, linked from *CBCNews.ca Home Page* at "Search cbc.ca" http://www.cbc.ca/canada/north/story/2008/11/28/nwest-vessel.html (accessed December 26, 2008).

⁶⁴ "Northwest Passage," linked from *geology.com Home Page* at "Geology Articles," http://geology.com/articles/northwest-passage.shtml (accessed December 26, 2008).

⁶⁵ Chris Mayer, "Northwest Passage Reopens Shipping Routes with Global Economic Impact," October 10, 2007, http://www.dailyreckoning.com.au/northwest-passage/2007/10/10/ (accessed February 6, 2009).

66. Defense Update – International Online Defense Magazine, 'Russia Extends its Arctic Naval Powerbase," http://www.defense-update.com/newscast/0808/070802_russian_navy_in_the_arcrtic.html (accessed December 27, 2008).

67. Bureau of Oceans and International Environmental and Scientific Affairs, "United States Climate Change Actions," December 2, 2008, http://www.state.gov/g/oes/rls/or/2008/112597.htm (accessed December 27, 2008).

68. *United Nations Climate Change UN Partnerships Home Page*, http://www.un.org/issues/ngo/n-clima.html (accessed December 27, 2008).

69. Bureau of Oceans and International Environmental and Scientific Affairs, "United States Climate Change."

70. U.S. Department of State Office of Oceans Affairs, "Arctic Policy," August 13, 2008, http://www.state.gov/g/oes/rls/rm/108788.htm (accessed December 27, 2008).

71. Ibid.

72. Ibid.

73. "Law of the Sea Briefing Book," October 29, 2007, linked from *Citizens for Global Solutions In the Beltway Home Page* at "The United States and the Law of the Sea: Time to Join," http://www.globalsolutions.org/in_the_beltway/united_states_and_law_sea_time_join (accessed December 28, 2008), 7.

74. Ibid., 7-8.

75. Carrie E. Donovan, "The Law of the Sea Treaty," April 2, 2004, http://www.heritage.org/Research/InternationalOrganizations/wm470.cfm (accessed February 6, 2009).

76. Citizens for Global Solutions In the Beltway, "Law of the Sea Briefing Book," 14-16.

77. John B. Bellinger III, "Generational Consistency in Law of the Sea," November 3, 2008, http://www.oceanlaw.org/index.php?name=News&file=article&sid=83 (accessed February 6, 2009).

78. Citizens for Global Solutions In the Beltway, "Law of the Sea Briefing Book," 7.

79. Bellinger, "Generational Consistency in Law of the Sea."

80. Citizens for Global Solutions In the Beltway, "Law of the Sea Briefing Book," 2.

81. Ibid., 8.

82. Coast Guard Office of Budget and Programs, *U.S. Coast Guard Posture Statement with 2009 Budget in Brief* (Washington, DC: U.S. Coast Guard, February 2009), 24.

83. Ibid., 33-34.

[84] Bureau of Oceans and International Environmental and Scientific Affairs, "United States Climate Change."

[85] United Nations, "UN's Kyoto Treaty Against Global Warming Comes into Force," February 16, 2005, http://www.un.org/apps/news/story.asp?NewsID=13359&Cr= global&Cr1=warm (accessed February 7, 2009).

[86] S. Fred Singer, "A Stronger Case Against Kyoto," May 14, 2001, http://www.independent.org/newsroom/article.asp?id=296 (accessed February 7, 2009).

[87] Sohbet Karbuz, "The US Military Oil Consumption," February 25, 2006, http://www.energybulletin.net/node/13199 (accessed December 28, 2008).

[88] Air Force Print News Today, "B-52 Flight uses Synthetic Fuel in all Eight Engines," December 15, 2006, http://www.af.mil/news/story_print.asp?id=123035568 (accessed December 28, 2008).

www.ingramcontent.com/pod-product-compliance
Lightning Source LLC
Chambersburg PA
CBHW081814170526
45167CB00008B/3443